Copyright © 2025 by Educate Learners

Published by Educate Learners

All rights reserved. No part of this publication may be reproduced, distributed, or transmitted in any form or by any means, including photocopying, recording, or other electronic or mechanical methods, without the prior written permission of the publisher, except in the case of brief quotations embodied in critical reviews and certain other noncommercial uses permitted by copyright law.

First Printing, 2025.

ISBN: 978-1-951573-64-5

www.educatelearners.com

It's One o'Clock

1:00

It's Two o'Clock

2:00

It's Three o'Clock

3:00

It's Four o'Clock

4:00

It's Five o'Clock

5:00

It's Six o'Clock

6:00

It's Seven o'Clock

7:00

It's Eight o'Clock

8:00

It's Nine o'Clock

9:00

It's Ten o'Clock

10:00

It's Eleven o'Clock

11:00

It's Twelve o'Clock

12:00

It's One Fifteen

1:15

It's Two Fifteen

2:15

It's Three Fifteen

3:15

It's Four Fifteen

4:15

It's Five Fifteen

5:15

It's Six Fifteen

6:15

It's Seven Fifteen

7:15

It's Eighteen Fifteen

8:15

It's Nine Fifteen

9:15

It's Ten Fifteen

10:15

It's Eleven Fifteen

11:15

It's Twelve Fifteen

12:15

It's One Thirty

1:30

It's Two Thirty

2:30

It's Three Thirty

3:30

It's Four Thirty

4:30

It's Five Thirty

5:30

It's Six Thirty

6:30

It's Seven Thirty

7:30

It's Eight Thirty

8:30

It's Nine Thirty

9:30

It's Ten Thirty

10:30

It's Eleven Thirty

11:30

It's Twelve Thirty

12:30

It's One Forty Five

1:45

It's Two Forty Five

2:45

It's Three Forty Five

3:45

It's Four Forty Five

4:45

It's Five Forty Five

5:45

It's Six Forty Five

6:45

It's Seven Forty Five

7:45

It's Eight Forty Five

8:45

It's Nine Forty Five

9:45

It's Ten Forty Five

10:45

It's Eleven Forty Five

11:45

It's Twelve Forty Five

12:45

Thank you for reading!

Get a free year long subscription to our online education resource library when you purchase any one of our books.

Code: EDBOOKS

educatelearners.com

www.ingramcontent.com/pod-product-compliance
Lightning Source LLC
Chambersburg PA
CBHW041602070526
44586CB00003BA/49